Northwestern

Wild Berries

J.E. Underhill

hancock
house

ISBN 0-88839-027-0
Copyright © 1980 J. E. Underhill

Second edition
09 08 07 06 05 04 03   10 9 8 7

Cataloging in Publication Data
Underhill, J. E., 1919–
  Northwestern wild berries.

  Bibliography: p.
  ISBN 0-88839-027-0

  1. Berries—Northwest, Pacific—Identification.
I. Title.
QK98.5.N6U53 1979   581.632   C79-091056-X

*We acknowledge the financial support of the Government of Canada through the
Book Publishing Industry Development Program (BPIDP) for our publishing activities.*

Published simultaneously in Canada and the United States by

HANCOCK HOUSE PUBLISHERS LTD.
19313 Zero Avenue, Surrey, B.C. V3S 9R9
(604) 538-1114 Fax (604) 538-2262

HANCOCK HOUSE PUBLISHERS
1431 Harrison Avenue, Blaine, WA 98230-5005
(604) 538-1114 Fax (604) 538-2262
Web Site: www.hancockhouse.com *email:* sales@hancockhouse.com

Contents

Photo Credits

Photos by the author, except for the following:
Ian G.Forbes: cover, pages 25, 25, 35, 45, 49, 56, 58, 61, 64, 80, 88.
Hilary Stewart: pages 47, 78.
R.W.Campbell: pages 20, 84.
Lewis Clark: pages 36, 85.
W.J.Merilees: pages 26, 33, 70, 71.
D.Hancock: pages 76, 92.
B.C.Parks Branch: pages 34, 37, 39, 60, 75, 82, 90.

Foreword

For many years it was my pleasant task to guide summer visitors through some of our British Columbia parks along woodland trails in the southern parts of the Province. These, to me, were specially happy summers, for the people I guided found so much obvious pleasure in the things we discovered together, and much of their pleasure rubbed off on me. It is to all of those people with whom I have shared the trails that I dedicate this book.

Berried plants seemed to be a subject of universal interest amongst the people who attended my walks and talks. A new berry quickly evoked the questions "Is it poisonous?" "Can I eat it?" and "Will you tell us how to use it?". It was this interest that led me, in 1967, to publish in pamphlet form the key presented in this book.

My time for writing is very limited, like that of any other working man with a family and friends. The book would likely still not be completed but for the help and encouragement of a number of good friends. I wish especially to thank David Stirling for reading my scribblings and making many good suggestions; Peg Martin and Betty Schenck for their skilled advice and my wife, Elise, for patiently enduring with me the pangs of my giving birth to a first book.

The leaf sketches and map that decorate the book are the work of Peter Eyles, one of B.C.'s up-and-coming young artists.

The colour photographs of berries are mostly from the collection of the B.C. Provincial Parks Branch, many of them have been taken by myself. Some of the very best, however, are the work of independent outside photographers. I gratefully acknowledge the contributions of Ian Forbes, Dr. Lewis Clark, William Merilees, and Wayne Campbell.

J.E. "Ted" Underhill

Introduction

I suppose I am a bit of a missionary about the many berried plants that are so easily found along our Pacific Northwest fence rows and forest edges. Certainly I've enjoyed many a memorable outdoor ramble by letting it centre around an exploration of the beauty of these plants, and my interest in which of them is tasty and wholesome, and which is to be shunned. This enjoyment seems greatest when I can share it with interested friends. I'm not a health food faddist, nor do I make the foods of the wild a significant part of my daily fare. They simply furnish trailside nibbles, an occasional pie or jar of jam, and a casual glass of wine. Nevertheless they have been a constant and infinitely pleasurable source of interest over the years. I invite you to share that interest in these pages.

Let us look together at the more commonly seen wild berried plants of the Pacific Northwest — roughly the area west of the Rocky Mountains from southern Alaska to the Columbia River or further. We won't cover every last kind, for some may be left out that few of us will ever see. This book is not for the serious botanist, but for the family or individual seeking simple and reliable information about the berries along our roads and trails.

Because this is not for the expert, we shall take some other liberties. We'll use the word "berry" in the popular sense, rather than in the more restricted sense applied by the botanist. Also, we shall lump together some closely related kinds of plants.

What is a "berry"? A definition that I like, though it's not mine, is that a berry "is a seed or seeds packaged in a moist, tasty pulp to encourage animals to eat it". I like this because it leads past what a berry is, and into "why" a berry is.

Think about the commonest and most successful of wild plants — things like Dandelions and Thistle and Broom — and you will easily recognize that each has some effective way of moving its seeds away from the parent plant, and away from each other. Dandelions and many others of the daisies have beautiful little parachutes that bear their seeds a considerable distance, even on a gentle breeze. Cottonwoods and Anemones, and many other quite unrelated things use variations of the same theme. Maples and Fir trees have seeds with wings that also ride the wind, though they cannot fly as far.

Brooms, Peas, Geraniums and many others don't wait for the wind, but have ingenious catapult systems that throw their seeds violently in every direction. With these as with Dandelions and others, the important thing is that much seed gets moved to where it at least has a chance of freedom from the competition of its own brothers and sisters and parents. With a little thought it is apparent that the further seeds can move the better will be their chance to colonize ground newly opened up by fire, logging, slides or stream action, — or even the few inches scratched clean by a grouse or a pocket gopher.

Still other plants have enlisted the involuntary aid of the animal kingdom. Cocklebur and Cheat Grass, for example, have seeds specially designed to hitch a ride on the fur — or clothing — of whatever animal brushes past.

But amongst the real experts in this business of moving about are the berries. They don't wait for the animal to come along — they invite it! They don't depend upon the whim of the wind, but upon the much greater carrying capacity of a bird or a chipmunk or a bear.

Some berry plants have gone so far along this line of development that their seeds have a covering that actually needs to pass through an animal's digestive system in order to be in best condition for germination. Whether or not it is carried by an animal, the berry plant seems to have an advantage in that its seeds reach the ground either with a handy supply of animal fertilizer, or with a convenient moist mass of fermenting fruit pulp.

Even the colours of most berries seem selected to bring them some advantage. Most are red or reddish, colours highly visible to birds and mammals that can carry the seed, but far less visible to insects which generally don't see reds well. This does not mean that the insects never find berries, of course, for some kinds of fruit are very apt to be "inhabited". In such cases it often happens that an animal eats the berry "inhabitant" and all. Sometimes any of us who eat wild berries acquire a little extra unexpected protein in our diet this way!

Birds are the greatest carriers of berry-wrapped seeds. Indeed the lives of berry-bearing plants and those of birds that eat them are often inextricably linked together. The plant provides the bird with berries as a seasonal food, with dense growth in which to hide for protection from predators, and with nest sites. The bird, for its part, looks after the important task of moving the plant's seeds to new locations, in some of which they may succeed. Where there are birds there are berries, and vice versa. Chipmunks, squirrels, skunks, raccoons, bears and others all do their bit at moving berries, but the birds are the principal carriers of most kinds.

Now that we are on the subject of animals and berries it is appropriate to introduce the matter of people and berries. The first question most people ask about a strange berry is "is it poisonous?" Perhaps it will trouble some of you that the answer cannot always be a clearcut "yes" or "no." There are a number of reasons for this cloudy state of affairs. Firstly, scientists simply haven't yet studied this aspect of some Pacific Northwest species; secondly, some species are somewhat toxic to a few sensitive people, or at certain stages of ripeness, or when grown in certain locations; and, thirdly, some kinds of otherwise perfectly wholesome fruit contain seeds that can cause poisoning if eaten to excess. Even the Apple, queen of our orchards, is guilty on this last count.

Still, the fact is clear that the Pacific Northwest has very few native berries that are likely to cause us trouble. One convenient contributing factor is that nature has endowed most of our non-poisonous kinds with flavours that are pleasant to most people, and our poisonous or suspect kinds with flavours that deter most people from eating enough to cause harm. Indeed, we can fairly safely say that those native berries that taste good to most people will be safe to eat. Nevertheless, it would be unwise to wander about sampling heavily on this basis.

One very important word of caution is necessary at this point. While our native berries are relatively safe in modest quantities, this is not so with the berries of the ornamental plants we grow in our gardens. Some of these are extremely poisonous, and a surprisingly high proportion of our garden favourites must be included in any "danger list" of things to be avoided as food.

Amongst poisonous garden berries the following are common in the Northwest, and are to be strictly avoided so far as tasting is concerned:

Holly (Ilex spp.)

Daphne (D. mezereum, D. laureola, & others)

Privet (Ligustrum vulgare)

Lily of the Valley (Convallaria majalis)

Yew (Taxus baccata and others)

It is not enough to know and avoid the berries of these plants as we find them in gardens. At least two of them, Holly and Spurge Laurel *(Daphne laureola)* have spread extensively into the woodlands surrounding our towns.

The sensible course is to avoid experimenting. Identify berries carefully, using the key and pictures that follow — and eat only those you are sure about!

One common fallacy is that we can safely eat whatever the birds or squirrels or other animals eat. This just isn't true at all. Birds, for example, eat quantities of Poison Ivy berries in the Interior, and at the coast they feast upon Holly berries — yet both of these are distinctly poisonous to us.

A curious thing about poisons in the plants we do eat is that quite a few of the fruits we prize most highly come from plants which have poisonous foliage or seeds. Apricots and some Cherries, for example, have leaves that are quite toxic, and have been known to kill livestock. Apple seeds are recorded as having killed at least one person who ate them in quantity. Probably it is wise to eat sparingly of the seeds of most fruits of the Rose Family — which includes Apricot, Peach, Apple, Plum, Cherry, and others, especially if they are uncooked.

But I am dwelling too long upon the thought of berries as something to eat — or to be poisoned by. They have other values, too. As has already been said, they attract birds and other animals whose presence we enjoy. But many berry plants have an even simpler virtue in just being extremely attractive to look at.

Who can drive the roads of the Pacific Northwest without noticing and enjoying the bright colour splashes of the Redberry and Blueberry Elders? Who can walk a mountain trail past carpets of scarlet Bunchberry without admiring the plant? Bright china-blue fruit of the Queen's Cup always attracts attention amidst the greens and browns of the forest floor. Red Huckleberries against the clean green of their foliage put me in mind of Christmas, even though I see them in August's heat. These and some others of our berried plants are just as showy as the Cotoneasters and Berberis that we grow in our gardens.

Look more closely, though, at some that are not so obviously showy, for these, too, often have beauty and interest that is only revealed to the discerning eye. A prime example is Black Twinberry. Stand back a dozen feet, and the whole plant seems a bit dowdy and uninteresting. Move in and examine it closely, and each pair of berries books like two glowing jet-black pearls held in a little cup of brilliant red. Twisted Stalk is another that belongs in this category. It doesn't command attention with a massive show of colour, but its ruby-red berries are so delicately poised on their curious kinked stalks, so nicely spaced along the branch, and of such a lovely colour that I think them to be one of the treats of the August forest.

While we are about the business of appreciating berried plants, it is a good time to think about their conservation. Throughout the more easily accessible parts of the Northwest we have left behind the days when the outdoors was so big and people so few that poor outdoors habits didn't show. Now there are so many of us eager for outdoor experiences that we must learn to tread lightly on our wild land, or in the very act of enjoying it we shall destroy it. In some of the most popular of our parks this is already being learned too late.

Speaking of parks, it should be pointed out that in most parks of the Pacific Northwest berry-picking is forbidden. There are two good reasons for this restriction: First, so many people flock to parks that they would soon trample these places to death if they engaged in activities such as berry-picking that took them off trails. Second, park berries serve a special purpose as specimen berries for all to enjoy seeing — rather than as goodies for a few to eat.

So plan to pick your berries outside of parks. This is no real hardship. It will give you an excuse to get a good map, and to begin exploring a lot of the little side roads you've been missing — a pursuit that can be good fun even without berries in mind.

One of the interesting aspects of the relationships between man and nature is that man, in the act of logging off the forests, has vastly increased the quantities of some kinds of berries. As we've said before, many berry plants are colonizers of newly opened land. Something similar has happened along the ditchbanks and hedgerows of our roads, where Wild Roses and Himalaya Berry thrive that could not have existed in the forests that stood there before the farmer or roadbuilder came.

Does berry-picking "interfere with nature"? Well, it is obvious that the berry you or I eat is one less for the Waxwing or the Chipmunk. However, the berry crop is normally many times greater than is necessary to supply the needs of these other animals. When the berries are scant they don't attract humans, and many kinds of berries don't attract us even if plentiful. Perhaps, if a very high proportion of people forsook the supermarkets and took to the hills for berries, then there might be some problems. I'll bet, though, that they'd give up and go home leaving ample for the other creatures. This means, too, that they would be leaving enough to permit the normal spread of the seed of these plants — the reason the berries exist in the first place. No, I don't think berry pickers pose a threat to nature, provided they are not concentrated in small areas like parks, and provided they exercise good conservation habits.

Possibly the reverse is true. I suspect that a lot of the families who pick berries acquire in the act a measure of knowledge of and respect for nature that leads them to be interested in protecting not just the berry patches, but all else in the outdoors.

When you or I do find a good berry patch, it definitely behooves us to think and do a bit of conservation. The basic thing is to avoid damage to the berry plants, and to all other growth around them. Above all, we should not break off branches to get at the fruit! We should leave no litter, for that is sheer laziness. We should pick a bush or a patch clean before moving on. That way we wear a minimum area, and we end up with more in the pot. People who snipe a few berries here and there trample much acreage, leave poor pickings for others, and end up with fewer berries for themselves!

So plan a few outings just to explore the world of the wild berries. Enjoy not only an exploration of their tastiness, fresh or cooked, but also open your eyes to their other values. You won't need to drive far from the city, for many berries thrive in empty suburban lots, and, indeed, you'll miss many as you drive that you'll see as you walk. Take this book along, and let it introduce you to the world of the plants with berries.

A simple

key to Wild Berries

The following pages contain a key to the identification of the more common and widespread kinds of wild berried plants of the Pacific Northwest. Some closely related kinds are lumped together.

For purposes of this key, trees are considered to be those plants that normally have woody stems (trunks) of over 2 inches diameter at 2 feet above the ground. It should be understood, however, that the distinction between trees and shrubs is not always clear-cut. The normally shrubby Service Berry may sometimes grow to tree-like proportions. Similarly, some people will prefer to consider Common Juniper as a tree, though we deal with it here as a shrub.

To identify a berried wild plant, examine it closely to determine three things:

    a) the general form of the plant
    b) how the berries are borne
    c) the shapes of the leaves

Then select consecutively the one number best describing your plant in each of the three key sections in turn. Use the resulting three-figure number to locate and identify your plant in the list that follows. Then proceed to the illustrated page dealing with the plant, and verify your identification using the photographs and text and sketch.

Remember — this key is not to be used for berried plants of gardens!

SECTION I (Choose the first of your three numbers here)

1. Tree — without spines or thorns on branches
2. Tree — with spines or thorns on branches
3. Upright shrub without spines or prickles on stems
4. Upright shrub with spines or prickles on stems
5. Prostrate mat-forming shrub
6. Trailing or climbing vine — without prickles
7. Trailing or climbing vine — with prickles on stems
8. Soft-stemmed plants — stems not woody

SECTION II (Choose the second of your three numbers here)

1. **Berries mostly borne singly.**

2. **Berries mostly in pairs or "Siamese twins".**

3. **Berries in seemingly stemless clusters.**

4. **Berries in obviously stemmed clusters.**

5. **Compound fruits with many seeds like those of raspberry and strawberry.**

14

SECTION III

(Choose the third of your three numbers here)

1. **Leaves borne single and margins not toothed.**

2. **Leaves borne single and margins finely to coarsely toothed.**

3. **Leaves lobed.**

4. **Leaves compound — divided into leaflets from a central point.**

5. **Leaves compound — divided into leaflets from a central stalk.**

6. **Leaves needle-like or scaly.**

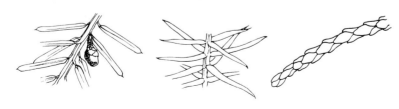

Plants with berries listed as "not recommended" are either suspected of causing trouble to some people who have eaten them, or there is lack of clear evidence of edibility, or, in a few cases, they are best left alone because they are easily confused with other, poisonous species.

No.	Ripe Berry Colour	Other Characteristics	Name	Edibility	Page No.
116	soft red	berry open one end	**WESTERN YEW**	DO NOT EAT	26
116	powder blue or greenish	evergreen tree	**ROCKY MTN JUNIPER**	edible	24
131 (151)	bright red	leaves opposite fruit hard	**PACIFIC DOGWOOD**	edible, not palatable	74
141 (142)	orange to red	coastal tree with papery red bark	**ARBUTUS**	edible	76
142 (132)	scarlet	leaves often spiney	**HOLLY**	POISONOUS	68
142 (342)	bright red to very dark red	berries glossy and smooth	**WILD CHERRY (3 kinds)**	edible, often bitter	50
142	purple/black	leaves strongly ribbed	**CASCARA**	edible	69
142	yellow to light red	some leaves lobed	**PACIFIC CRABAPPLE**	edible	53
145 (345)	scarlet	"ladder" leaves	**MOUNTAIN ASH**	edible but mealy	63
151	bright red	leaves opposite fruit hard	**PACIFIC DOGWOOD**	see 131	74
212	blue/black with powdery surface	may resemble erect shrub	**SLOE**	edible	52
242	dull purple/ black	thorns sparse but very long	**BLACK HAWTHORN**	edible	46
243	dull red	berries persist to late winter	**EUROPEAN HAWTHORN**	edible	48

311 (312)	red, blue or blue/black	leaves of some kinds minutely toothed	**HUCKLE-BERRIES BLUEBERRIES**	edible and choice	82
313 (343)	various	dried flower often hangs on fruit	**WILD CURRANT (several)**	edible	40
316 (416)	powdery blue/green	evergreen shrub	**COMMON JUNIPER**	edible	24
321	dull red	"Siamese twin fruit" east of Cascades	**RED TWINBERRY**	edible	88
321	shiny black	bright red calyx cup behind berry pairs	**BLACK TWINBERRY**	not recommended	87
331	bright red	cinnamon spots on leaf bottoms	**SOOPOLALLIE**	edible	70
331	waxy white	berries persist into winter	**WAXBERRY**	not recommended	91
341	blue/black	fairly tall shrub	**INDIAN PLUM**	edible	50
341	white to pale blue	twigs quite red	**RED-OSIER DOGWOOD**	edible, not palatable	75
341	reddish	much branched shrub — red papery bark	**MANZANITA**	edible	78
341	greenish white	leaves in threes	**POISON IVY**	see 344	66
342	dull purple black	berries not hairy leaves thin, soft	**SERVICE BERRY**	edible	44
342	dull purple black	berries hairy leaves thick, stiff — berry stalks sticky	**SALAL**	edible	80
343	shiny red	tall slender shrub leaves mostly 4" wide, or more	**SQUASH-BERRY**	edible	92
343	various	broad low shrubs with smaller and often hairy leaves	**WILD CURRANT**	see 313	40
344 (341)	pale green/white	leaves in threes	**POISON IVY POISON OAK**	POISONOUS POISONOUS	66

345	dark red	berries velvet hairy	**SUMACH**	edible	65
345	shiny bright red	berries ⅛″ dia. in big bunches	**RED-BERRY ELDER**	not recommended	90
345	powdery blue	big shrub, some-times treelike	**BLUEBERRY ELDER**	edible	88
345	powdery blue	low shrub, stiff "holly" leaves	**OREGON GRAPE**	edible	38
353	dull red	big maple-like hairy leaves	**THIMBLE-BERRY**	edible	62
416	blue/green	prickly spread-ing evergreen	**COMMON JUNIPER**	edible	24
443	bright red	huge leaves-very prickly stems	**DEVIL'S CLUB**	not recommended	72
443	various	leaves under 4″ broad	**WILD GOOSE-BERRY (several)**	edible	42
445 (415)	bright red	fruit hard	**WILD ROSE**	edible	54
454 (455)	yellow, red, or reddish black	fruit separates from a fleshy cone	**SALMONBERRY WILD RASPBERRY BLACKCAP**	all edible and choice	58 60 60
511	red	a bog plant	**CRANBERRY**	edible	84
511	red	not a bog plant leaves oval	**MOUNTAIN CRANBERRY**	edible and choice	84
511 (541)	red	not a bog plant lvs. broadest beyond the middle	**KINNIKINNICK**	edible	78
522 (511)	red to blackish	leaves very broad	**ALPINE WINTER-GREEN — SLENDER WINTERGREEN**	edible edible	80
512	white	very tiny leaves	**CREEPING SNOWBERRY**	edible	81
516	glossy black	heather-like	**CROWBERRY**	edible	64
631	orange to red	berries in leafy green "cup"	**HONEYSUCKLE**	not recommended	86
641 (643)	bright red	variable annual vine, purple flowers	**BITTERSWEET**	POISONOUS	85

754 **(755)**	red to blackish	thorny vines	**BLACKBERRY** **(several)**	edible and choice	**56**
811	shiny blue	low woodland plant two broad leaves	**QUEEN'S CUP**	not palatable	**28**
811	red	leaves broad, veins parallel, fruit often in pairs	**FAIRY BELLS** **(several)**	edible	**30**
811	red	broad leaves, parallel veins, fruit pendant in a row	**TWISTED STALK**	edible	**32**
811	orange to scarlet	leaves narrow, pointed	**BASTARD** **TOAD-FLAX**	not recommended	**36**
816	bright red	to 6' tall, leaves very fine	**ASPARAGUS**	not recommended	**27**
831	bright red	low plant, often forms carpets	**BUNCHBERRY**	edible	**72**
841	mottled green then red	24" tall, leaves parallel veined	**FALSE** **SOLOMON'S** **SEAL**	not recommended	**34**
841	red	about 10" tall, leaves parallel — veined	**WILD LILY-OF-** **THE VALLEY**	not recommended	**34**
842	shiny purple/ black	purple flowers	**BLACK** **NIGHTSHADE**	POISONOUS (see text)	**85**
843 **(844)**	red or white (2 forms)	berries held stiffly from central stalk	**BANEBERRY**	POISONOUS	**37**
844 **(845)**	dark purple	leaves in 3's or 5's evenly-toothed edges	**WILD** **SARSAPARILLA**	edible	**71**
854	red	leaves blue/hairy	**WILD** **STRAWBERRY**	edible and choice	**49**

Berries

on the Bush

Some sunny summer weekend, pack the family, this book, and some pots and pans into your car, and do an outing especially for the purpose of getting acquainted with the wild berries. Visit a variety of places, — the ditchbanks and hedge-rows amongst farm fields of the lowlands; the logged-over lower slopes of the hills; and the timbered forests wherever they occur.

Fairly surely you will find and identify some berry kinds you hadn't known before. Perhaps, too, you'll bring home the berries for a pie or shortcake or other culinary treat. Even better, you will have had some healthy exercise out in the fresh air, and you will have made the acquaintance, not only of plants with berries, but also of a host of flowering plants, birds, insects and other animals. Some of these may lead you to explore other places and other books, for there is no end to the beauty and interest of this out-of-doors that lies all about us in the Pacific Northwest.

On the pages that follow, the plant species, or groups of species are set out in their natural or phylogenetic order, so as to bring related kinds together.

Names at the tops of the pages refer to the species illustrated, other related species are often referred to in the body of the text. Especially in the cases of larger plant groups, no attempt has been made to list more than a few representative or distinctive species.

Wild berries grow almost everywhere, from dry interior valleys, to wet coastal rainforest; from alpine tundra to sea level.

COMMON JUNIPER *(Juniperus communis L.)*

ROCKY MTN. JUNIPER *(Juniperus scopulorum Sarg.)*

Juniper berries very commonly evoke thoughts of gin. True enough, it is the berries of a European Juniper, very like Common Juniper, along with various herbs, that give gin its flavour and bouquet.

We actually have three kinds of Junipers that are widespread in our area. The two named above are illustrated here, and are easily distinguished. Common Juniper has needle-like leaves in whorls of threes, and is a spreading shrub. Rocky Mountain Juniper is tree-like, and has fine scaly foliage. The third, Creeping Juniper, *(Juniperus horizontalis* Moench), has foliage like that of Rocky Mountain Juniper, but is a trailing shrub.

Common Juniper   EDIBLE (SEE TEXT)

In all three of these Junipers the fruit, though berry-like, is actually a small cone in which the scales are fused together. In colour the fruit is variable, usually being greenish, but covered with a bluish bloom.

Junipers are characteristically plants of dry places, especially favouring dry rock bluffs. There, where they are exposed to high winds, they often assume grotesque and sometimes beautiful shapes.

I can't imagine any reason for gathering juniper berries in quantity, but every other year or so I like to gather and dry a few handsful. A very few of these crushed into a bread and onion stuffing for Cornish Rock Fowl — or for a roasting chicken — provide a tasty flavour that is different.

Rocky Mtn. Juniper   EDIBLE (SEE TEXT)

WESTERN YEW *(Taxus brevifolia* Nutt)

Centuries ago, while English bowmen were learning to use their English Yew *(Taxus baccata)* to make their famous longbows, Indians here were putting Western Yew to similar use. Yews of several species grow widely in temperate regions of the Northern Hemisphere, and northern man has long valued its beautiful supple wood.

Western Yew is a small tree of moist but well drained ground over much of our area. The bark is notably red and papery, while the needles are sharp-tipped but very soft, and are borne flat in two rows. I once heard Western Yew described as "a hybrid between Douglas Fir and the Arbutus"! The person who said that was a long way out in scientific fact, but I liked her description of Yew's appearance just the same.

In summer the Yew trees may carry berry-like red fruit. These are best left alone. Other Yew species are notoriously poisonous, containing a powerful alkaloid heart depressant which can cause death with very few advance symptoms. This is contained in most parts of the plant. Our Yew has been less well studied, and so far is thought to be less dangerous. Nevertheless, there remains the risk that some strains may be toxic, or that one will pick the wrong kind of Yew berry. I would strongly advise leaving them alone, despite their attractive flavour, and despite the fact that I have myself tried the pulp (not the seeds) without ill effect.

Western Yew   DO NOT EAT

ASPARAGUS *(Asparagus officinalis* L.)

If sexuality is the popular topic of our age, then Asparagus should be a popular plant to write about! In case you haven't noticed it already, there are separate male and female Asparagus plants — the female, of course, bearing the berries.

Asparagus has been with us a long long time, having been harvested by the early Romans and Greeks from the European and Asiatic shores of the Mediterranean several thousand years ago. As we know it it is the result of development by European growers. Settlers brought it to our shores a century or more ago, and it has since escaped cultivation. It especially favours moderately watered but well drained soils of the interior of our area.

City folk, accustomed to Asparagus as they see it in a can or on supermarket shelves are apt not to recognize the full-grown plant. It is four feet or more in height, and branches out to form a wispy mass of fine lacy foliage.

Be warned that the tender green shoots that taste so good when cooked, can cause a nasty skin irritation if nibbled raw. The bright red berries look inviting, but they, too, contain irritating principals, and should be left alone.

Asparagus   NOT RECOMMENDED

QUEEN'S CUP (*Clintonia uniflora* (Schult.) Kunth.)

In the Pacific Northwest, a lone bright blue berry borne erect between two attractive simple leaves can only be Queen's Cup.

Like so many other plants of the Lily Family, Queen's Cup also has handsome flowers. These are pure white, six-parted, and well over an inch across. Watch for them in moist forests at medium elevations from the coast to the Rockies. Often you'll find them in association with the Bunchberry.

The blue berries are too scarce, and of too poor a flavour to interest berrypickers. Also, in my opinion, they are much too attractive to be picked! It is recorded that grouse sometimes feed on them.

The Latin name, *Clintonia*, commemorates Governor De Witt Clinton (1769-1828) of New York State. Amongst his other accomplishments, the Governor was a botanist of considerable ability.

Queen's Cup   NOT PALATABLE

HOOKER'S FAIRY BELL *(Disporum hookeri* (Torr.) Nicholson)

ROUGH-FRUITED FAIRY BELL *(Disporum trachycarpum* (Wats.) B&H.)

Fairy bells grace moist forest floors throughout most of our range. In general appearance they fairly closely resemble the related Twisted Stalks. Both have handsome broad clasping leaves positioned alternately up the stalk. In Fairy Bells, however, the flowers and berries are borne at the tips of the branches in scant clusters, while in Twisted Stalk they spring solitary from each leaf axil to form a row up the stalk.

Hooker's Fairy Bell has leaves with long tapering points, and it bears smooth berries. Rough-Fruited Fairy Bell, as its name implies, has a distinctly roughish berry, and the broad leaves have rather abrupt short tips. In both the flowers are about half inch across, and are whitish to pale yellow green, and tend to be almost hidden in the foliage.

Berries of Fairy Bells are brilliant red when ripe, and they are quite large — about half inch in diameter. They are edible, with a sweetish, bland, and rather insipid flavour. Indians used to gather them along with everything else that was edible, but today they are left for the Chipmunks and Ruffed Grouse.

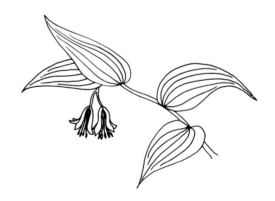

Hooker's Fairy Bell   EDIBLE

Rough-Fruited Fairy Bell   EDIBLE

31

TWISTED STALK *(Streptopus amplexifolius* (L) DC)

SIMPLE-STEMMED TWISTED STALK *(Streptopus roseus* Michx.)

Lilies are mostly woodland plants, and Twisted Stalk is no exception. We may watch for it in moist forests over much of North America.

Examine the flower stalk to see the distinct sharp kink that gives the plant its name. This, too, clearly sets it apart from two other lilies it resembles, Fairy Bells and False Solomon's Seal.

Twisted Stalk's oval red berries are amongst the handsomest ornaments of the forest. Seen by itself, the individual berry isn't at all special, but they hang evenly-spaced in a graceful arching row. This, set off by the lovely broad leaves, give a very fine total effect.

The berries are edible, and were used to some extent by the Indians, but to us the flavour is poor, sweetish, bland and insipid. Amongst wild animals, the grouse and chipmunks are principal foragers of this fruit.

Twisted Stalk   EDIBLE

Simple-Stemmed Twisted Stalk has decorative bell-shaped rose-coloured flowers, and is a less common and much smaller plant. Even smaller and rarer is Small Twisted Stalk *(Streptopus streptoides* (Ledeb.) Frye & Rigg.). Its few rose-coloured flowers are broadly open with recurved petals.

Simple-Stemmed Twisted Stalk   EDIBLE

FALSE SOLOMON'S SEAL *(Smilacina racemosa* (L.) Desf.)

WILD LILY-OF-THE-VALLEY *(Maianthemum dilatatum* (Wood) Nels. & MacBr.)

False Solomon's Seal grows to about three feet high, almost always in clumps of several graceful arching stems. Each stem is well clad with broad parallel-veined leaves that clasp the stem, and each is topped with a much-branched panicle of many tiny flowers. Later come the round berries, which are at first mottled green and red, then become reddish.

Also common here is the smaller eighteen-inch Star-Flowered False Solomon's Seal *(Smilacina stellata* (L.) Desf.) which has much narrower leaves and fewer, larger flowers in an un-branched raceme. The few berries are reddish to dark blue.

False Solomon's SEAL   NOT RECOMMENDED

Wild Lily-Of-The-Valley is much shorter, being up to about twelve inches tall. It has more-or-less heart-shaped leaves, and distinct leaf stalks. The berries are pale red.

None of these berries can be classed as poisonous, though there is evidence that, eaten raw in quantity, they act as a strong purgative. Indians did use them, but presumably not in quantity, for none of them are easily gathered in great abundance.

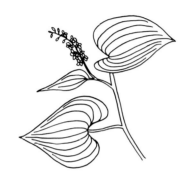

Wild Lily-of-the-Valley   NOT RECOMMENDED

BASTARD TOAD-FLAX (*Comandra livida* Richards)

There just has to be an interesting story behind a name like that, but I'm afraid I can only decipher part of it. The real Toad-Flax is Linaria, so that the name must stem, in part, from a fancied resemblance between Comandra and Linaria.

There are two Comandras in our area. *Comandra livida*, illustrated here, grows to about 12″ tall, and bears a few — usually three — scarlet berries in the axils of the upper leaves. *Comandra umbellata* (L.) Nutt.) grows further south in Washington and Oregon, and bears bluish to brownish berries in an umbel above the leaves.

Animals appear to harvest berries of these plants freely, but not much is known about their safety for humans. Most authors who mention these plants at all consider the fruit edible but of rather worthless quality. There is some evidence, however, that C. *umbellata* may concentrate poisonous selenium in its fruit. Until more is known they should probably be sampled sparingly.

Bastard Toad-Flax   NOT RECOMMENDED

BANEBERRY *(Actaea rubra* (Ait.) Willd.)

Baneberry is probably as poisonous a fruit as the Northwest can muster up. Certainly it is a berry that you shouldn't eat in any quantity, but you won't drop dead of trying one berry. Actually it is interesting to do this — just to satisfy one's curiousity. One needn't, and shouldn't, swallow the thing.

That Baneberry is indeed poisonous seems well documented. It is listed as a "violent irritating purgative and emetic" by one reputable authority. Being of an inquisitive, though not very daring mind, I've sampled and spat out a berry or two. I experienced no ill effects, but found the flavour very acrid and disagreeable. I have no inclination to experiment further. It is said that half-a-dozen berries will cause strong symptoms in an adult. Baneberry belongs to the Buttercup Family — a group that contains enough poisonous plants to encourage me to leave it alone.

Most Baneberry plants in our region have bright Chinese red berries, but on a small percentage of plants they mature snow white. Often the two colour forms grow side by side. Both are highly decorative.

Baneberry — POISONOUS

OREGON GRAPE *(Berberis nervosa* Pursh)

Oregon Grape is rather like the poet — unsung in its own country. We take it pretty much for granted, along with the other plants of our woods. Elsewhere it receives much more recognition. David Douglas, the Scottish botanist who roamed these parts in the 1830's quickly recognized it as a fine ornamental shrub. To this day it is highly regarded in the gardens of some far distant countries.

We actually have three species. One is low growing *(Berberis repens* Lindl.); the second, about 18 inches tall, is Oregon's state flower; and the third *(Berberis aquifolium* Pursh), grows to three or four feet in height. All have similarly attractive sprays of bright yellow spring flowers, followed by masses of powder-blue berries. To whomever gave the plant its common name, these must have resembled bunches of grapes, though there is no real relationship.

Oregon Grape

Many people are very fond of a tart Oregon Grape jelly as a condiment to use with meats. I've never cared much for it. When my mother made it, years ago, we kids dubbed it "quinine jelly", and we rolled on the floor in simulated agony! Maybe Mom didn't have the right recipe.

Still other people are devotees of Oregon Grape wine. I've made it myself, and I've tasted that made by others. Its big drawback is a pronounced earthy taste that takes at least a year or more to age out. It seems to me that there are so many other fruits that yield better wine so much more easily!

Oregon Grape   EDIBLE

FLOWERING CURRANT *(Ribes sanguineum* Pursh.)

SQUAW CURRANT *(Ribes cereum* Dougl.)

Currants are simply Gooseberries without prickles. We have about a dozen kinds of wild Currants in our area, and as many Gooseberries. All are moderate-sized shrubs with "maple-like" lobed leaves. Our Gooseberries are described on the next page.

Probably the best known of our Currants is the Flowering Currant of the western slopes. This has striking sprays of deep pink flowers, and is, indeed, a flowering shrub of real garden value. It goes rather unrecognized here, probably just because we take it for granted as a "native weed". With a little pruning after flowering it can be shaped to suit the gardener, and it is well worth growing. The berries are pale lavender and not of interest for eating.

Flowering Currant   EDIBLE

Fruits of other wild Currants are diverse in colour, ranging from greenish through blackish to some with blue or purple bloom. None are so tasty, and few are sufficiently abundant, to attract berrypickers. Squaw Currant of dry interior hillsides is sometimes abundant and attractive with its scarlet berries, but no one that I know of seems to have developed a taste for its rather acrid and bitter fruit.

In balance, this is a group that we should probably best enjoy for its beauty of flower and form. Sometimes it is nice to look at a berry without thinking of it as something to eat!

Flowering Currant

Squaw Currant   EDIBLE

SWAMP GOOSEBERRY *(Ribes lacustre* (Pers.) Poir.)

(AND OTHERS)

Gooseberries have prickles or spines, Currants do not. Apart from that they are all members of the same group, the genus *Ribes,* and have but slight differences.

Here we deal with the spiny Gooseberries, while the Currants appear on the previous page. In our area there are about a dozen kinds of Gooseberries, most looking fairly similar to one another.

Swamp Gooseberry, one of the commonest, turns up along stream banks and swamp margins over a wide area. It is a shrub about three to five feet tall, carrying curious pendant flowers in June, followed by small mahogany-red or purplish berries in late summer. This is one of those plants that has much charm, but which must be examined closely to be appreciated.

Swamp Gooseberry   EDIBLE

The fresh berries don't have what I consider to be a very pleasant flavour, and they are difficult to pick in quantity. Nevertheless, I once gathered a few pounds and made them up into a wine. It turned out to be one of the finest wild berry wines I have ever made.

Swamp Gooseberry Flowers

Berry Country in Autumn

SERVICE BERRY *(Amelanchier alnifolia* Nutt.)

(SASKATOON BERRY)

This berry is quite differently regarded in different parts of its range. Undoubtedly, this is partly a matter of there being several different strains of the plant, each in its own area. Probably, too, local climate has much effect.

To people from east of the Rockies, this berry is highly regarded as the plump sweet basis for fine pies and jams. In the dry interior of the Pacific Northwest it is often abundant, and sometimes bears heavily, yet doesn't seem to attract very many human customers. Bears and Chipmunks often have the crop almost to themselves. The plant is quite common right out to the coast, but fruits sparsely there, and ripens unevenly.

Give the plant its due, though, for it had its place in history. On the Great Plains this was the principal fruit pounded up with buffalo meat to make the famous pemmican — the chewy "iron ration" that sustained many a voyageur through his rigorous western journeys.

Service Berry Flowers

David Thompson's narrative describes it thus: "Pemican, a wholesome, well tasted nutritious food — is made of the lean parts of the Bison dried, smoked, and pounded fine; in this state it is called Beat Meat. — Pimmecan is made up of — twenty pounds of soft and the same of hard fat, slowly melted together, — and poured on fifty pounds of Beat Meat, well mixed together and closely packed. — on the Great Plains there is a shrub bearing a very sweet berry of a dark colour, much sought after, great quantities are dried by the Natives — and as much as possible mixed to make Pemmecan." (three spelling's of Pemican are those of Thompson's.)

We have tried our west slope Saskatoons for winemaking, and for jams and pies, but can only conclude that it must be inferior to the crop our Prairie cousins praise with such enthusiasm.

Service Berry   EDIBLE

BLACK HAWTHORN *(Crataegus douglasii* Lindl.)

COLUMBIA HAWTHORN *(Crataegus columbiana* Howell)

Black Hawthorn is native to streamsides and brush thickets throughout much of the Pacific Northwest. Its fierce, long, sharp thorns identify it at any season, but its purplish-black berries make it distinctive in late summer.

As with other kinds of Hawthorns from further east, the fruit varies somewhat from tree to tree in colour, size, and flavour. Unfortunately it never seems to rise above a very poor standard in edibility. The flavour of the ripe fruit is rather bland, slightly sweet, and a bit "dry and puckery". Native Indians once used it, but it finds little favour today.

Black Hawthorn   EDIBLE

Black Hawthorn reminds me of an amusing incident: I once joined the nature walk of a young park naturalist who introduced his audience to a Black Hawthorn, and had each person sample a berry or two. When he moved on down the trail, one tourist, obviously intrigued by the excellent presentation, stayed behind to sample another berry and to explore inside two or three. Catching up with the group at the next stop he created quite a furor when he announced "Mr. Naturalist, did you know that each of those hawthorn berries is full of little white worms?!"

Columbia Hawthorn has red fruit, longer thorns, and a more easterly distribution in our range. Otherwise it is fairly similar to Black Hawthorn.

Columbia Hawthorn   EDIBLE

EUROPEAN RED HAWTHORN *(Crataegus monogyna* Jacq.)

This is the familiar "Maytree" of our boulevards and gardens, a close relative of our native Black Hawthorn. In parts of the Pacific Northwest the European Hawthorn has long ago escaped cultivation, and has spread far and wide along ditch-banks and into waste fields. Undoubtedly the birds have had a hand in this, for they make much use of the Hawthorn "haws" when earlier foods are gone and winter snows make other foraging difficult.

From a man-oriented point of view, I'm inclined to think that the highest virtue of the fruit of the Hawthorn is its decorative appearance. The bright red to dark mahogany berries hang on the trees far into winter when other bright colour has gone, and furnish welcome relief to an otherwise bleak scene.

I can't get excited about the culinary possibilities of this fruit. I've spent some hours emulating Euell Gibbons by sampl-ing fruit from many trees. The flavour and colour vary, but the flavour never rises to the point of being really inviting. At best it is faintly apple-like, with overtones of bitterness. The rest is mealy nothingness.

Nevertheless, a wine made from European Hawthorn reveals hidden virtues. It yields a rich rose of considerable quality. For winemaking the plant offers the advantages that it is avail-able in quantity, it is quick to pick, and it can be picked on a mild day in December or January when other fruit is not available.

European Red Hawthorn   EDIBLE

WILD STRAWBERRY *(Fragaria* spp.)

Isaac Walton, author of the "Compleat Angler", penned the lines "Doubtless God could have made a better berry than the Strawberry, but doubtless God never did!" He was referring to the little wild Strawberry, for the big ones of our gardens had not then been developed.

The Pacific Northwest has several species, all fairly similar except to the botanist. A coastal species that ranges from here far down the west of South America is one of the parents of our cultivated kinds of Strawberries. (The other parent was an Eastern North American wild Strawberry). Thomas Laxton was the hybridist, and his production was the great "British Sovereign" variety that is still widely grown.

This was an especially interesting horticultural development. Nearly all of our cultivated fruit varieties have been with man for many centuries, with but slow improvement. Cultivated Strawberries are the one really widely grown fruit crop that has originated within the past century.

With all due respect to Thomas Laxton and his "British Sovereign" and other garden varieties, none yet have the marvellous flavour of the little wild berries that still grow unchanged on our rock bluffs and roadsides!

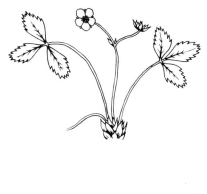

Wild Strawberry   EDIBLE AND CHOICE

49

INDIAN PLUM *(Osmaronia cerasiformis* (T&G) Greene)

Earliest of all the Northwest Pacific Coast shrubs to bloom in spring is the Indian Plum. The flowers are little pendants of white and cream — inconspicuous from a distance, but attractive if examined closely.

This is another of the group of plants that contains the cherries, plums, apricots and peaches, but beside these tasty relatives its fruit is a disappointment to taste. It is safe enough, but has a rather unpleasant flat and "puckery" flavour.

But fruit doesn't have to be for us to eat to bring us a benefit. Indian Plum is avidly sought out by the birds — so avidly, in fact, that they seldom let it ripen.

Indian Plum in Flower

Indian Plum　　EDIBLE

CHOKECHERRY *(Prunus virginiana* L.)

BITTER CHERRY *(P. emarginata* (Dougl.) Walpers)

Three kinds of wild cherries occur in our area — Chokecherry, Pin Cherry, and Bitter Cherry. All have typical cherry leaves — smooth, finely toothed, and with sharp tips. Chokecherry and Bitter Cherry characteristically also bear a pair of tiny raised "glands" at the base of each leaf. Chokecherry, with its heavy trusses of mahogany red berries is the plant most people know, especially in the dry interior valleys where it thrives. It is most often tree-like, but sometimes bears fruit while still just a shrub by our artificial standards.

Chokecherries occur across the continent, and early attracted the attention of the settlers. One of these wrote in 1634 that: "chokecherries so furre the mouthe that the tongue will cleave to the roofe, and the throate wax hoarse!" A taste of the raw fruit will quickly convince you that three hundred years or so have not changed the ways of the Chokecherry!

Chokecherries do make a very palatable, if distinctive, wine. Also they may be crushed and boiled, and the juice extracted to make a very nice jelly. Old timers claim that the trick is to let the fruit stay on the tree as long as possible. Novices pucker up and shudder!

If you sample any of the cherries raw — including cultivated kinds, — avoid eating the seeds. These contain amygdalin, a substance that breaks down in the body to yield deadly cyanide.

Bitter Cherry

Chokecherry

Chokecherry   EDIBLE

Bitter Cherry   EDIBLE

SLOE *(Prunus spinosa L.)*

There's nothing slow about the sloe once it is growing in a spot it likes! The sloe, a kind of ornamental plum, was introduced into this area many years ago as a small garden tree — valued both for its abundant white blossoms and for its powder-blue fruit. Since then, perhaps with an assist from the local birds, it has spread out sporadically into our hedgerows and ditchbanks.

In spring the big clusters of white flowers remind one of the Hawthorn, to which Sloe is related. The fruit, however, is very different, being a cherry-sized bluish plum — quite attractive to look at, but intensely acid to taste.

Sloe is perhaps best known as the basis for sloe gin — not really a gin at all, but a fruit liqueur, made by steeping sloe fruit several months in white brandy. A very acceptable substitute may be made at home by covering sloe berries with vodka — pricking the fruit so as to release its flavour (but do not mash and create a cloudiness). After a couple of months the liquid is strained off and sweetened to taste with sugar — and used sparingly!

Less well known is the strong likelihood that the Sloe in centuries past became one of the parents of some of the cultivated varieties of plums that are grown in today's orchards.

Sloe   EDIBLE

PACIFIC CRABAPPLE *(Pyrus fusca* Raf.)

Pacific Crabapple is a small tree of moist valley bottoms along the coast and west of the Cascade Range from Alaska to Northern California. Most people never notice it, for in every way it is a thoroughly inconspicuous tree.

Early Indians knew it well, however, and made much use of its pale bronze little oval "apples". These they gathered in August, storing them for winter in bags woven from cattails. Sometimes the crabapples were mashed with Salal berries or other fruit.

From the Indians the early white settlers learned to use this fruit, but they made it into jelly in the European style.

Few people use the Wild Crabapple today, but there is still plenty of it around for those who wish to experiment with something different in the way of pie or jelly or applesauce. The seeds should be strained out, however, for, like other apple seeds, they can be dangerous if eaten in quantity — especially when raw.

Pacific Crabapple   EDIBLE

WILD ROSE *(Rosa* spp.)

Most of the berries we like come rushing into ripeness in mid to late summer, then are quickly gone. For a time there seem just too many of them to be coped with.

This is why we are grateful to the Wild Roses. True, their bright red "hips" are in evidence in August, but we may be content to simply admire their handsomeness then. They won't rush away. A fine weekend in October is time for a rose hip foray, and we've even found them in usable condition in January in some winters. A few frosts seem to improve their flavour, and are reputed to increase their remarkably high content of vitamin C.

The biggest chore in using rose hips is that of getting rid of the mass of hard hairy seeds, for it is only the shell of the hip that we want. The chore seems a little easier if the hips are allowed to dry a day or two before being slit open. It is also said that the seeds are more easily removed from summer-picked hips, but I'm of the opinion that they are not then at their best for flavour.

Wild Rose in Flower

Rose hips are an admirable and highly nutritious food that may be used in a variety of ways — even chewed raw. My wife makes a fine syrup for hotcakes or waffles by boiling the hips and pouring the strained extract onto sugar. A few handsful of rose hips add a piquant flavour if added to other fruits being fermented for wine.

Wild Rose "Hips"   EDIBLE

PACIFIC BLACKBERRY *(Rubus ursinus* Cham. & Schlecht.)

(TRAILING BLACKBERRY)

HIMALAYA BERRY *(Rubus procerus* Muell.)

(HIMALAYAN BLACKBERRY)

EVERGREEN BLACKBERRY *(Rubus laciniatus* Willd.)

In August of every year there is a period of two or three weeks when my arms and legs are as scratched as if I had taken up Cougar wrestling! This is Blackberry season, when the luscious black fruit is ripe, and my wife and I fill our pails with the makings of many a fine jam and pie, and of many gallons of good wine.

Like most people, we gather lots of Himalaya Berries, largely because they are abundant in the suburbs near where we live, but also because they are good for so many uses. The origin of the name puzzled me a long time, but I found its story in one of Luther Burbank's books. Evidently he developed this berry from stock sent to him from India, introducing and naming it in 1885. Today, of course, it is probably the easiest to find of all our wild-growing berries.

Pacific Blackberry   EDIBLE

56

With or near Himalaya Berry you may likely find the Evergreen Blackberry. It has a similar growth habit, but the leaves are quite finely cut, and the taste, to my mind is very inferior.

A berry superior to either of the above for flavour is the Pacific Blackberry, a true native of these parts. It is also called the Dewberry. The best place to find it is in logged-over land a few years after the loggers have moved out, for it is very much a plant colonizer.

Himalaya Berry   EDIBLE AND CHOICE

Evergreen Blackberry   EDIBLE

SALMONBERRY *(Rubus spectabilis* Pursh.)

Salmonberry — now there's a strange name! How do you relate a fish to a berry?

Well, like so many other things of the Pacific Northwest, this stems from our native Indians. They, being a perceptive people, and close to nature, noticed the similarity between the little globules of the berry and the eggs of the salmon that teemed in the streams beside which the berry grew. Perhaps they even fancied that there might be some relationship.

Indians gathered the fruit, as they gathered virtually all edible wild berries. Also they harvested the tender young shoots of Salmonberry in spring. This was "muck-a-muck", eaten fresh or roasted, and regarded as a spring treat.

Salmonberry   EDIBLE

Salmonberry, because the fruit slips off the cone that carries it, is grouped with the wild Raspberries. As we might expect of such a rampant stream-side plant, the fruit tends to be watery and somewhat insipid, and is usually borne rather sparsely. Still, it makes a refreshing nibble on a hot day's hike. The berries are sometimes yellow, sometimes red, with the former seemingly having the better flavour.

Along west coast trails the Salmonberry and Blueberry or Huckleberry often grow not far apart. A bowl of the two mixed with cream and sugar makes a tasty way to start the day.

Salmonberry flower

WILD RED RASPBERRY *(Rubus idaeus* L.)

BLACKCAP *(Rubus leucodermis* Dougl.)

CREEPING RASPBERRY *(Rubus pedatus* J.E. Smith)

Raspberries and Blackberries are all kinds of *Rubus,* the artificial distinction being made that, in the case of the Raspberries, the berry slips loose from its cone-like receptacle. In the Blackberries the receptacle stays within the berry.

Wild Red Raspberry is widespread and locally abundant in our area, especially on rock slides and bluffs of the middle elevations in the interior. Generally their flavour is excellent, and a bowl of them with rich cream makes any meal a feast for the Gods. Red Raspberry in extremely closely related form grows around the world in the northern hemisphere, and has given rise to our garden strains.

Wild Red Raspberry   EDIBLE AND CHOICE

Blackcap has a fairly similar range. It is distinguished by the pale bluish colour of its canes, and by the berries, which are dome-shaped and almost black when ripe. Eaten fresh they are a little dry, but they make truly excellent jam and jelly. Wine made from this berry has won several competitions in this area. It matures quickly, and displays a simply fantastic bouquet.

Creeping Raspberry makes a pretty ground-covering mat in moist spruce forests of our Pacific Northwest, from the seacoast well up into the mountains. Both the dainty white flowers and the Chinese red berries are highly decorative. The more northern Cloudberry *(Rubus chamaemorus* L.) has yellowish berries. *R. lasiococcus* Gray is like Creeping Raspberry, but has 3 lobed rather than 5 lobed leaves.

Creeping Raspberry
EDIBLE

Blackcap   EDIBLE AND CHOICE

THIMBLEBERRY *(Rubus parviflorus* Nutt.)

This, being one of the commonest berried plants of roadsides in the moist parts of the Pacific Northwest, is bound to attract some attention. Partly this is because of its prominent big maple-like leaves. Seldom, though, do people seem to notice Thimbleberry for its dome-shaped fruit — actually a kind of Raspberry. Unlike the Raspberry, this plant bears no prickles.

In my estimation the Thimbleberry is most valuable for its softly decorative leaves and large white flowers in early summer. The fruit is uneven in ripening, hard to gather in quantity, and rather uninteresting in flavour. At best it makes an acceptable "padding" for a bowl of something more tasty — perhaps Raspberries or Blue Huckleberries.

Thimbleberry stalks are often attacked by tiny insects that lay their eggs within the stalks. The eggs hatch into little grubs that feed within the stalks, setting up irritation. The irritation causes the plant to grow a large swollen gall, and may make the stem assume grotesque and curious shapes. Flower arrangers sometimes gather these galled stalks in winter, and may value them highly.

Thimbleberry   EDIBLE

SITKA MOUNTAIN ASH *(Sorbus sitchensis* Roemer)

Mountain Ash is not an Ash at all, for Ash belongs to the Pea Family and Mountain Ash to the Rose Family. The similarity in names arises because the leaves happen to have some resemblance. Mountain Ashes of a number of species grow around the world in the northern hemisphere. Best known here is the European Mountain Ash *(Sorbus aucuparia* L.) that is so often grown on our city boulevards.

Not many people are really aware that we have our own native kinds. Superficially there is much resemblance, though the European species grows into a much larger tree. The easiest way to tell them apart is that the European plant has 11 to 15 leaflets in the compound leaf, while Sitka Mountain Ash has 9 to 11. Some of our native plants also develop a deeper and more beautiful berry colour. Why they are not more widely cultivated I do not know. Ours also put on a truly spectacular, if brief, display of autumn leaf colour.

Man doesn't make much use of the mealy, bitter and acrid fruit of Mountain Ash, but it is obviously much favoured by the birds. They seem to take it after fermentation has started, and often appear to get a bit giddy. Are we observing avian alcoholism?

In Scotland the European Mountain Ash is known as the Rowan, and its berries are valued for wine-making. I don't know anyone here who has tried using this fruit, but if it produces as good a wine as the closely-related European Hawthorn then it is well worth gathering.

Sitka Mountain Ash   EDIBLE

CROWBERRY *(Empetrum nigrum* L.)

Crowberry is unknown to most people of the southern parts of the Pacific Northwest, even though it occurs fairly widely in this region. For the most part it grows in swamps or high in the hills, so escapes the attention of people. In northern British Columbia and Alaska the plant is far better known.

To some northern Indian and Eskimo groups the fruit of the Crowberry was — and indeed still is — an important item of diet.

When you first chance upon a plant of Crowberry you'll perhaps think you've found a heather, for it has similar small evergreen leaves, and the same low shrubby growth. Despite the resemblance, Crowberry is not really closely related to the heathers. Crowberry flowers are dark and quite inconspicuous. The berries, about ¼" in diameter, are glossy black, and are borne down amongst the tangle of the foliage.

To some people the flavour of Crowberry takes some getting used to, but to most it is soon very acceptable. The berries make fine jams and jellies.

Crowberry   EDIBLE

64

SUMACH *(Rhus glabra* L.)

Sumach is a common native shrub of the dry interior valleys, but often turns up as an ornamental in coastal gardens, where it is valued for its interesting summer foliage and brilliant red autumn colours. In the wild it makes a fine autumn companion for the bronzy shades of the Oaks, or for the bright yellow blossoms of Rabbitbush.

Sumach is closely related to Poison Ivy, with which it sometimes grows, and to the eastern Poison Sumach. Fortunately it doesn't share its cousins' unfriendly attitude towards mankind. It lacks the poisonous oil, urushiol, that makes them such a hazard.

In early summer each Sumach stalk grows a broad mop of attractive compound leaves. In the centre of this, later in summer, is the conical cluster of tiny, dark red berries. These are velvety-hairy, the hairs containing an acid that has a flavour very reminiscent of lemons. Indians used to steep the crushed berries in water to make a refreshing drink — rather like our lemonade. I've heard it said that the sieved pulp of the berries can be used to flavour a custard for a sort of "lemon pie". I haven't tried this, but it sounds reasonable if I may judge from the taste of the berries.

Sumach in Autumn

Sumach   EDIBLE

65

POISON IVY *(Rhus radicans* L.)

POISON OAK *(Rhus diversiloba* T&G)

Poison Ivy is common in dry interior valleys from central B.C. southwards through our area, and ranges from there east to the Atlantic coast. Poison Oak is coastal from the Gulf Islands southwards. Both are plants everyone should know, for they, above all other plants in North America, have the ability to cause human suffering.

The key thing to look for is *irregularly-toothed leaves borne in threes.* Poison Ivy is, in our area, a low shrub with sharp-pointed leaflets. Poison Oak grows three or four feet tall, and has rounded leaflets. Both bear dingy greenish-white berries, and both put on a fine show of leaf colour in autumn.

Both cause us misery with a skin-poisoning resin that in most people causes severe irritation.

Berries of these plants are definitely not for human consumption. I once met a young boy who had tried some. His mother reported that he became very ill, and vomited violently, but recovered without apparent lasting effects. There have been reports of worse cases. It is interesting to note that these same berries are a favoured food for many kinds of birds. There's your answer for anyone who claims that man can safely eat what the birds eat!

Poison Ivy flower

Poison Ivy in Autumn

Poison Ivy — POISONOUS

ENGLISH HOLLY *(Ilex aquifolium)*

Holly is quite rapidly spreading out from gardens and commercial plantings into surrounding forests of the moister parts of B.C., Washington, and Oregon. Birds, especially Robins, have distributed the seed far and wide.

Many of the plants of Holly you encounter in the forest bear no fruit. There are separate male and female plants in Holly, the males, of course, carrying only the pollen. The leaves are most often spiney, but smooth-edged kinds are found as well. Some, undoubtedly, are hybrids, for there are many kinds of Holly.

Holly, and its bright red berries, have been with man since before history began. The ancient pagan Romans used to celebrate the festival of Saturnalia, honouring Saturn, god of the harvest, about Dec. 21st. of each year. For the occasion they used much Holly as decoration, and gave it to one another as a token of goodwill. When Christianity came along and introduced Christmas, the custom of using Holly easily transferred to the new festive date.

The name Holly comes from an old English word "holem", the meaning of which is obscure. Apparently, though, it is not at all connected with the word "holy".

Holly berries are quite poisonous to humans, causing nausea, vomiting, diarrhea, and possible death if ingested in quantity. Birds, however, eat them with great gusto, and no apparent ill effects.

English Holly — POISONOUS

CASCARA *(Rhamnus purshiana* D.C.)

Cascara grows along the Pacific slope from British Columbia into California. It is one of the "Buckthorns", a group of trees and shrubs growing around the world in the northern hemisphere. Several of them are of interest for their garden beauty or their medicinal properties.

Our Cascara is, of course, best known for the medicinal use of its bark as a laxative. This is one of the things the white man early learned from the Indian, who had from long ago steeped the pounded bark in cold water to make his medicine. Most of us who grew up prior to World War II have indelible memories of the efficiency of this old Indian remedy.

The berries are edible, but of poor quality. There is no record that they share the powerful properties of the bark. Birds take them quite avidly in late summer and autumn when other food becomes scarce. They are a blue-black colour when ripe, but ripen unevenly, so that a single bunch will often show a spectrum of colours, green through yellow to reddish and almost black.

Cascara thrives almost unnoticed today, but twenty-five years ago its bark brought such a price that it was an endangered plant, and controls had to be instituted to protect it.

Cascara   EDIBLE

SOOPOLLALIE *(Shepherdia canadensis* (L.) Nutt.)

Soopollalie is one of the commonest of our berries, yet one of the least known. Probably this is because it just doesn't please most of our palates, but it wasn't always that way. This was a fruit that was highly prized by the early Indians. Indeed, some of them still use it.

Early Indians gathered the fruit in midsummer, crushing it with a little water, then whipping it to a froth with a whisk of twigs. The foam was eaten with a special flat wooden spoon, and was often mixed with fish oil. Today's Indians are more apt to mix the froth with sugar, and I've heard them refer to it as "Indian ice cream".

There seems to be a distinct case here of differing ethnic tastes. While most Indians apparently relish this fruit, to most Caucasians it tastes decidedly unpalatable. One acquaintance likens it to a mixture of soap, vinegar and turpentine!

Soopollalie turns up throughout our area on dryish, well drained soils, but is especially common in the pine forests of medium elevations in the Interior.

Look closely at Soopollalie and notice that there are separate male and female plants, only the female, of course, bearing the fruit. Another interesting feature is the tiny patches of cinnamon-coloured hairs on the lower surfaces of the leaves. Under a magnifying glass these may be seen to be star-shaped and really quite handsome.

Soopollalie   EDIBLE

WILD SARSAPARILLA *(Aralia nudicaulis* L.)

The name Sarsaparilla is Spanish in origin, deriving from "zarza" — bramble, and "parilla" — the diminutive of "parra" or vine. The true Sarsaparilla flavouring was obtained from the roots of any of several tropical American plants of the genus Smilax. In its day this was a very popular flavouring, but it has passed out of vogue.

Our wild Sarsaparilla has somewhat similar flavour and medicinal qualities in its roots, so gained some use as a substitute. Today the plant passes virtually unnoticed, though often abundant, in the semi-moist interior forests. Sometimes those who do notice it confuse it with Poison Ivy, for it often bears leaves in threes. The resemblance stops there, however, for Wild Sarsaparilla has leaf edges that are evenly toothed, while those of Poison Ivy have but a few coarse irregular teeth.

The flowers and berries, too, are inconspicuous. The flowers are small and dingy-white, rather like those of the Ivy, while the berries are like raisins. Both tend to be lost in the leaves. The fruit is edible, but not of any interest.

Wild Sarsaparilla   EDIBLE

71

DEVIL'S CLUB *(Oplopanax horridum* (J.E. Smith) Miq.)

Devil's Club is the bane of loggers, surveyors, prospectors and others who must bushwhack the moist valleys of our western slopes. Each long sprawling stem is abundantly armed with vicious sharp spines that some people claim are poisonous. To make matters worse, adjoining shrubs are often entangled, and present such a barrier that the toughest woodsman is glad to make a detour.

Each long stem is topped in summer with a few great big leaves, maple-like in shape, and often over twenty inches broad. At the heart of the leaves, in August, stands a single conical cluster of tiny scarlet berries. Often the cluster droops over to one side if the crop is heavy.

This is another of the kinds of berries about which little or nothing is known as far as toxicity is concerned. That being the case, they are best left alone. I've tasted a few, and find them distinctly acrid and unpleasant.

Devil's Club   NOT RECOMMENDED

BUNCHBERRY *(Cornus canadensis* L.)

Bunchberry, for which many prefer the name Dwarf Dogwood, is one of the fairest and most widespread of our dwarf forest plants. It ranges from Alaska to New Mexico, and from the Pacific to the Atlantic.

Bunchberry usually grows about 6 to 8 inches tall, and spreads underground to form dense mats, often many feet across. Near the top of each little upright stem is one main whorl-like group of attractive leaves. Above these is borne what looks like one large 4-petalled flower. On closer inspection it turns out, however, that the white "petals" are really modified leaves (bracts), and that the real flowers are a group of tiny purplish things clustered at the centre of the four bracts.

Late summer visitors to the forests will find the white "flowers" replaced by clusters of brilliant coral-red fruits that give the plant its name. These "berries" are quite edible, but are mealy and a little bitter. Nevertheless some Indian tribes gathered them in quantity to eat, and others have used them with some success in puddings. Grouse eat considerable quantities.

To make its decorative value complete, Bunchberry's leaves turn brilliant shades of bronze, red and purple before dropping in the autumn.

Bunchberry

Bunchberry   EDIBLE

73

PACIFIC DOGWOOD *(Cornus nuttallii* Aud. ex T&G)

(FLOWERING DOGWOOD)

When in bloom, the Pacific Dogwood is the most spectacular of all wild trees in our range. The "blossoms" are about 4 inches across, and are often borne in great abundance, especially if the tree is growing in the open.

As in the case of its tiny relative the Bunchberry, what appear to be 4 or 5 big white petals are actually floral leaves (bracts). The flowers themselves are tiny and inconspicuous, forming a tight greenish or purplish button-like cluster in the middle.

Pacific Dogwoods occur from S.W. British Columbia southward into California west of and in the Cascades, and are found in Idaho. In British Columbia the species has been adopted as the Provincial floral emblem, and it is strictly protected by law from all picking or other disturbance.

The brilliant red fruits are so jammed into their tight cluster that each is compressed into an angular form by its neighbours. They are edible, but hard and mealy and rather bitter, and fit only for emergency use by humans.

Pacific Dogwood in Bloom

Pacific Dogwood   EDIBLE, NOT PALATABLE

RED-OSIER DOGWOOD *(Cornus stolonifera* Michx.)

Red-Osier Dogwood is prolific along stream banks and around ponds and marshes from the coast to quite high up in the mountains. The name derives from its bright red twigs, which are most conspicuous in winter when the shrub is naked of leaves.

Some coastal Indian groups have long recognized the medicinal value of this bright red bark as a purge, making an extract in hot water. Some are said to still use it today before important canoe races.

The tiny white flowers are borne in large flattish "nosegays". Only a botanist would recognize their relationship to the Bunchberry and to the Pacific Flowering Dogwood.

The berries, also in broad clusters, vary from whitish to greenish blue. The birds seek it eagerly, so that it isn't often seen ripe in great quantity. I have no record of the berries being used by anyone in these parts, but a lady from northern Europe once told me that in her native land a very similar species produces a good country wine.

Red-Osier Dogwood   EDIBLE, NOT PALATABLE

75

ARBUTUS *(Arbutus menziesii* Pursh.)

The Arbutus or "Madrona" is a tree of the dry rock bluffs overlooking the Pacific in the narrow "Mediterranean" climate zone that extends from about Campbell River southwards along the Gulf of Georgia, down Puget Sound and the west coast of Washington and Oregon, and on through California. Always, in our area, you will find Arbutus near enough to the sea so that it may enjoy moist salt air, but always on gravel or rock, where its toes will stay dry. Truly, this is a fussy plant.

Arbutus has the distinction of being one of the handsomest of our native trees. Look at its graceful twisted trunks, its papery bright red bark; the big glossy evergreen leaves; the creamy flower trusses in early summer; and the attractive orange-red berries in the fall.

Arbutus

These last, however, are berries we may admire for their handsomeness alone. They are thoroughly mealy, bland, and a little bitter, and we find no record of Indian or white man being tempted by them, though they are apparently not poisonous. Birds, however, do use them avidly as late winter food.

Arbutus   EDIBLE

On the Trail in Berry Country

KINNIKINNICK *(Arctostaphylos uva-ursi* (L.) Spreng.)
(BEARBERRY)

MANZANITA *(Arctostaphylos columbiana* Piper)

There's no mistaking the Kinnikinnick, for it is a very distinct-ive ground-hugging, mat forming shrub, with leathery spoon-shaped leaves and red berries. In our area only the edible Mountain Cranberry of the north is at all similar.

Kinnikinnick, like most other plants, is particular as to where it grows. Look beneath it, and you will find rock, or coarse gravels, or sand.

Kinnikinnick Flowers

Kinnikinnick   EDIBLE

The name is Indian, and is said to mean "something to smoke". Indians used to dry and crush the leaves, smoking them alone or mixed with tobacco. Another name for this plant is Bearberry.

Indians also used the dry mealy berries, storing them to use in winter, for this fruit keeps better than most. The berries are said to make a refreshing drink obtained by boiling them, straining off the juice, and allowing it to settle, then sweetening to taste.

Two closely related plants grow in the far north of our area. These are the Alpine Bearberry, with shiny black fruit, and the Red Bearberry, which looks fairly similar to Kinnikinnick. Both are edible.

Manzanita is a decorative, much-branched shrub of coastal rock bluffs. It has reddish, papery bark similar to that of Arbutus, and bears reddish berries. It often hybridizes with Kinnikinnick where the two grow together, the hybrids being intermediate in appearance.

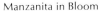
Manzanita in Bloom

SALAL *(Gaultheria shallon Pursh.)*

SLENDER WINTERGREEN *(Gaultheria ovatifolia Gray)*

Salal is the "forgotten fruit" of the Pacific Northwest. Few people use it, despite the fact that it is enormously abundant. Perhaps this is because the berries eaten out-of-hand are not as attractive as the Blackberries and Blueberries that most people seek.

Things weren't always this way. Indians of our Pacific Coast were very aware of the virtues of Salal, and ate great quantities. Often they mashed the berries, and dried the pulp on mats of Skunk Cabbage leaves. The dried pulp cakes were then rolled up and stored for winter use. Later they might be served with oolichan grease as a nutritious delicacy.

Slender Wintergreen   EDIBLE

Today, those of us who really know Salal find it a plant well worth seeking out in late August when the berries hang plump and soft. A jelly from these is so exceptionally good that some day it is bound to attract commercial attention. It has a sort of "supergrape" aroma. The fruit is easy to pick, for one uses the whole cluster, stem and all — and it jells easily.

Winemakers, too, will find Salal useful. Pure Salal wine lacks flavour, but has an intense aroma. A few ounces of it blended into wine from another fruit that lacks aroma can do wonders. I've never tried mixing the fruit at the outset, but this should work equally well.

There are two miniature Salals, Alpine Wintergreen (*Gaultheria humifusa* (Grah.) Rydb.), and Creeping Snowberry (*G. hispidula* (L.) Muhl.) with very tiny leaves and snow white berries that taste of wintergreen.

Salal Flowers

Salal   EDIBLE

HUCKLEBERRIES, *(Vaccinium* Spp.)

BLUEBERRIES

What is it — Blueberry — Blue Huckleberry — or just Huckleberry? There's a question that is very apt to raise an argument! The fact is that we have about a dozen kinds of related shrubs in the genus Vaccinium that all bear some claim to one or more of these names. Each has its universally recognized Latin name, but the common names tend to be localized and confused. To make matters more complicated, easterners use the name "Huckleberry" for an entirely unrelated plant.

But why argue? Our Huckleberries, Blueberries, Billberries, Whortleberries or Cranberries or whatever else you choose to call them, are all delectable additions to the outdoor larder. Black bears think so, too, and are often encountered harvesting this fruit.

Red Huckleberry   EDIBLE AND CHOICE

The commonest kind in our lowlands is the Red Huckleberry. In good crop years it is well worth seeking out, for its bright red berries make fine pies and jellies, and are delicious cooked into muffins or pancakes.

All the Vacciniums, whatever the colour of their berries, seem to produce good country wines, though there is evidence that it is sometimes hard to get them to start fermenting.

Blueberry   EDIBLE AND CHOICE

BOG CRANBERRY *(Vaccinium oxycoccus* L. var *intermedium* Gray)

MOUNTAIN CRANBERRY *(V. vitis-idae* L. ssp *minus* (Lodd.) Hult.)

These, our native Cranberries, are both closely related to the Cranberry of commerce, *V. macrocarpon,* that graces the Christmas turkey. Our native kinds, however, bear smaller fruit. Both, too, being Vacciniums, are closely related to the Blueberries and Huckleberries discussed on another page.

Bog Cranberry, and a very similar northern species, are slender little trailing plants that sprawl on the damp sphagnum of bogs. Their alternate leaves are shiny dark green above and silver-grey beneath, and are quite small. The flowers, too, are tiny, 4-parted, and to me rather resemble miniature fuchsias. The berries are white and red varying to bright red, and close to the size of a commercial Cranberry.

Mountain Cranberry is a mat-forming prostrate shrub of drier sites from bog margins into the forests. Superficially it much resembles the Kinnikinnick with which it sometimes grows, but has dark spots on the lower surfaces of its oval leaves, and has berries that are thin-skinned and juicy. Kinnikinnick lacks spots on its more "spoon-like" leaves, and has berries that are thick-skinned and mealy.

Our wild Cranberries make excellent substitutes for the kind commercially grown, and may be used to make pies and jams as well. Their flavour raw gives little hint of their potential when cooked

Mountain Cranberry   EDIBLE

BITTERSWEET *(Solanum dulcamara* L.)

BLACK NIGHTSHADE *(Solanum nigrum* L.)

Bittersweet and Black Nightshade are two closely related plants of the Potatoe Family, a family notorious for its poisonous properties. Did you know that the green tissue of potatoes is poisonous, and that people have succumbed from drinking the "tea" from boiled tomato leaves?

Bittersweet is a sprawling soft vine, with some leaves that are lobed at their bases. Its berries are bright red at maturity, and it usually bears a few flowers along with fruit at all stages of ripeness. Black Nightshade grows erect, has some leaves with coarse teeth, and bears berries that are blackish when ripe. Both are plants that prefer moist soil in open places.

Both contain in their fruit the poisonous alkaloid solanine, and each contains other, separate, poisons. Either can cause quite severe symptoms if eaten in quantity, especially if raw or unripe. In Black Nightshade especially, the toxicity apparently decreases markedly as the fruit ripens. Some people use its ripe berries for pies without apparent ill effect, and there are garden strains grown for their fruit.

Bittersweet — POISONOUS   Black Nightshade — POISONOUS

HONEYSUCKLE *(Lonicera ciliosa* (Pursh) DC)

Widespread along forest edges of the Pacific Northwest is this trailing or climbing vine. Most people first notice it because of its striking flowers. These occur as a cluster of elongated 1¼ inch trumpets of the brightest flaming orange colour. We have no other flower with which to confuse it.

The broad, rounded, blue-green leaves are quietly attractive in their own right, those behind each bunch of flowers being fused to form a beautiful shallow "cup". It is in this cup, that the fruit appears — a seemingly stemless cluster of berries that mature to a reddish colour.

As in the case of the closely allied bush Honeysuckles, these berries are neither poisonous nor palatable. The few that I have tasted had a strong, acrid and very disagreeable flavour. This quite obviously doesn't discourage the birds and bears and chipmunks, who harvest them with apparent enjoyment.

Within our area are several other vine Honeysuckle kinds with similar berries, but sporting more quietly coloured flowers.

Honeysuckle

Honeysuckle   NOT RECOMMENDED

BLACK TWINBERRY *(Lonicera involucrata* (Rich.) Banks ex Spreng.)

Black Twinberry is much more widespread than its relative the Red Twinberry, but it still is not a plant familiar to many people. Perhaps this is because its dark blackish fruit, borne in pairs in a bright crimson calyx cup look rather saturnine and dangerous to many.

The early Indians must have felt the same way — or maybe they had more positive knowledge. In any case some Indian groups claimed that eating these berries would "make one crazy". A search of available literature doesn't reveal medical records to back this claim, but I can vouch for the fact that the vile flavour by itself would be sufficient to drive one "around the bend".

In any case the early Indians were usually pretty sound in their judgement as to what was or was not good to eat. We would be wise to consider Black Twinberry as something to be looked at rather than as food.

Black Twinberry Flowers

Black Twinberry   NOT RECOMMENDED

RED TWINBERRY *(Lonicera utahensis* Wats.)

Red Twinberry grows east of the summit of the Cascade Range, so is marginal to the range of this book, and scarcely familiar to most people of the Pacific Northwest.

The plant is a rather open, but neatly-shaped shrub of up to about four feet in height, with broad glaucous blue-green leaves, pairs of yellow, trumpet-shaped flowers, and red fruit in distinctive "Siamese twin" pairs. It is actually one of the bush Honeysuckles, though, apart from the leaves, one would not readily spot its relationship with the vine Honeysuckles most of us are more familiar with.

The berries are borne sparingly, so are not apt to attract the attention of pickers. There seems no record of use by the Indians, and I've not heard of anyone else doing more than casually sampling the fruit. Little is recorded as to edibility or toxicity, but I have sampled quite a few with no apparent ill effects, finding them sweetish, but insipid.

Red Twinberry   EDIBLE

BLUE-BERRY ELDER *(Sambucus cerulea* Raf.)

Blue-berry Elder commands attention! This is a big shrub — sometimes quite tree-like, with large compound pale green leaves. It bears great flat-topped bouquets of creamy white flowers, and tops off the season with striking big clusters of tiny powder-blue berries.

This species occurs here and there along the coast, but is most frequently encountered in the warm interior valleys, where it seems to especially thrive in the disturbed soil of the roadbanks.

The berries are sufficiently harmless so that they may be used in a variety of ways, especially if cooked, but other parts of the plant are poisonous to eat. It is reported that the berries may cause nausea if eaten raw in abundance. Still, lots of people use and enjoy Blue-berry Elder fruit in jams, jellies and wines. I've used it for wine, and it is not bad, but tends to be rather high in astringent tannin.

Many people who try Blue-berry Elder for winemaking are probably attracted by the reputation of the somewhat similar European Elder. I understand that this is very variable, but that better forms are much superior to our local species. European Elder makes wine of such good quality that in by-gone days unscrupulous merchants passed it off as grape wine. Some countries eventually passed laws forbidding its use.

Blue-Berry Elder Blossoms

Blue-Berry Elder   EDIBLE

RED-BERRY ELDER *(Sambucus racemosa* L.)

Red-berry Elder is a joy to behold in August, with its great trusses of brilliant scarlet berries. This is one of our commonest shrubs, thousands of them growing along the roadsides of coastal valleys, and even well up into the mountains.

This is definitely one of the kinds of berries to be enjoyed for its eye appeal rather than for any taste appeal. Go ahead and taste a few — then you'll quickly understand what I mean!

There may be more to be avoided than just bad flavour, however, for there is some evidence that Red-berry Elder has made some people ill when consumed in quantity. There seems to be no record of the Indians eating them, and that in itself is likely a warning, for the Indians would scarcely have passed up so abundant a food without good reason.

The birds obviously find no reason to hesitate about Red-berry Elder. A number of kinds of birds — among them Swainson's Thrush and the Band-tailed Pigeon — scarcely wait for these berries to ripen before engaging in a greedy orgy.

Red-Berry Elder

Red-Berry Elder   NOT RECOMMENDED

WAXBERRY *(Symphoricarpus albus* (L.) Blake)

Waxberry is a plant we really notice in late autumn and early winter, for it is abundant over much of our area, and the large waxy white berries are conspicuous against the brown hues of the season. The rest of the year its soft blue-green leaves and little pink bell-like flowers go almost unnoticed.

This is one of those berries about whom different authors seem to disagree. Some list it as poisonous, while others consider it safe to eat but not palatable. The few berries I've nibbled at tasted awful — perhaps rather less so after frost. Probably the truth lies somewhere in the middle. There are four species in our area, and it is entirely possible that they differ in edibility while looking very similar to most people. Also there may be a situation where edibility differs with location. All of this is purely academic from my viewpoint, since I still maintain that the berry tastes awful. I'm quite content to leave this fruit to the Band-tailed Pigeons who sometimes take it as a late winter food around our place.

It is recorded that the Indians rubbed the berries on skin sores or burns, or rashes. They, like me, apparently shunned this berry as a food.

Waxberry   NOT RECOMMENDED

SQUASHBERRY *(Viburnum edule* (Michx.) Raf.)

HIGHBUSH CRANBERRY *(Viburnum opulusL.)*

Squashberry is a tallish shrub that often forms dense thickets along river bottoms and swamp margins. It is easiest to spot in autumn, when its three-lobed leaves glow a deep rich red, highlighted by the brighter scarlet berries.

The berries are easily told by their large flattish seeds and broadly flattened shape. These, of course, give the plant its common name. Raw, the fruit is decidedly tart and bitter, though it sweetens somewhat after a goodly touch of frost. It is recorded that some early West Coast Indians relished squashberries cooked with oil, but they attract extremely few of us today.

My only major personal experiment with squashberry was in winemaking. There it was an utter disaster! The finished product, even at a year's maturity, had a peculiar disagreeable flavour so bad that I gladly consigned it to the toilet. That, in our house, is a rare fate for any wine!

Squashberry   EDIBLE

Highbush Cranberry, like Squashberry, is a kind of Viburnum — closely related to several popular garden Viburnums. Unlike the Squashberry, it reveals the relationship in its large handsome flower clusters. Each of these consists of a central cluster of insignificant petal-less blossoms, with an outer ring of large white petalled blooms.

The berries are tart when raw, but when cooked with sugar make a sauce fairly closely resembling that from the quite unrelated Cranberry.

Highbush Cranberry   EDIBLE

Bibliography

For those who would like to read for themselves some of the background information upon which this book is based, the following are some useful references:

ANDERSON, E.	Plants, Man, and Life. Univ. of Calif. Press. 1967.
BAILEY, Liberty Hyde	A Sketch of the Evolution of Our Native Fruits. MacMillan. 1898.
BROWN, Annora	Old Man's Garden. Gray Publishers. 1970.
CRAIGHEAD, J.J.	
CRAIGHEAD, F.C.	Field Guide to Rocky Mountain Wildflowers.
DAVIS, R.S.	Peterson Field Guide Series No. 14. 1963.
DOMICO, Terry	Wild Harvest. Hancock House. 1979.
FYLES, Faith	Principal Poisonous Plants of Canada. King's Printer, Ottawa. 1920.
GIBBONS, Euell	Stalking the Wild Asparagus. 1962. Stalking the Blue-Eyed Scallop. 1964. Stalking the Healthful Herbs. 1966. David McKay Co. Inc.
HITCHCOCK, C. Leo	
CRONQUIST, Arthur	
OWNBEY, Marion	Vascular Plants of the Pacific Northwest
THOMPSON, J.W.	5 volumes 1955-69. Univ. of Wash. Press.
HUME, H. Harold	Hollies. MacMillan. 1953.
JAMES, Wilma R.	Know Your Poisonous Plants. Naturegraph Publishers. 1973.
KINGSBURY, John M.	Poisonous Plants of the United States and Canada Prentice Hall. 1964. Deadly Harvest. Allen & Unwin. 1972.
KIRK, Donald R.	Wild Edible Plants of Western United States. Naturegraph Publishers. 1970.
MUENSCHER, W.C.L.	Poisonous Plants of the United States. MacMillan. 1951.
NELSON, Alexander	Medical Botany. Livingstone. 1951.
SZCZAWINSKI, A.F.	The Heather Family of British Columbia. B.C. Prov. Museum Handbook No. 19. 1962.
SZCZAWINSKI, A.F.	Guide to Common Edible Plants of B.C.
HARDY, G.A.	B.C. Prov. Museum Handbook No. 20. 1962.
TAYLOR, T.M.C.	The Rose Family of British Columbia. B.C. Prov. Museum Handbook No. 30. 1973. The Lily Family of British Columbia. B.C. Prov. Museum Handbook No. 25. 1966.
VIERECK, L.A.	Alaska Trees and Shrubs
LITTLE, E.L. Jr.	USDA Handbook No. 410. 1972.

Index